IMAGES
of America
READING FIREFIGHTING

On the cover: Please see page 34. (Courtesy of the Reading Fire Department.)

IMAGES of America
READING FIREFIGHTING

Michael J. Kitsock and Michael R. Glore
for the Reading Fire Department
Foreword by Chief William Rehr

ARCADIA
PUBLISHING

Copyright © 2008 by Michael J. Kitsock and Michael R. Glore
ISBN 978-1-5316-3508-4

Published by Arcadia Publishing
Charleston SC, Chicago IL, Portsmouth NH, San Francisco CA

Library of Congress Catalog Card Number: 2007931354

For all general information contact Arcadia Publishing at:
Telephone 843-853-2070
Fax 843-853-0044
E-mail sales@arcadiapublishing.com
For customer service and orders:
Toll-Free 1-888-313-2665

Visit us on the Internet at www.arcadiapublishing.com

The Firemen's Monument was erected in City Park and dedicated September 2, 1901. It was originally located across from the Berks County prison at Columbus Drive and Reservoir Road, but was moved to a more prominent place at the head of Penn Street in the 1980s when the roadway connecting Penn Street to Reservoir Road was closed and landscaped.

Contents

Acknowledgments 6

Foreword 7

1. Reading Towne Organizes a Fire Department 9

2. The Department Motorization 31

3. Modern Era 65

Acknowledgments

In no fire department across the United States was the role of the volunteer firefighter more greatly extolled than in Reading, Pennsylvania. With over 10,000 volunteer firefighters manning 14 fire stations and 30 pieces of fire apparatus, the Reading Fire Department boasted the largest all-volunteer department in the United States. From major floods, great influenza epidemics, world wars, and major conflagrations, this fire department served with distinction and honor since Colonial times. *Reading Firefighting* commemorates over 200 years of firefighting history in the city of Reading.

Reading Firefighting would not be possible without the assistance and archival information of the Historical Society of Berks County and the Reading Fire Department. Their photographs captured the history and spirit of the Reading Fire Department in action from its early existence to present times. Special thanks are also accorded to Anthony Miccicke, Reading Fire Department historian, who has steadfastly chronicled the history of the Reading Fire Department for many years. Anthony currently serves as a dispatcher for the City of Reading.

The greatest supporter and contributor to *Reading Firefighting* has been Reading fire chief William (Bill) Rehr. Bill has contributed numerous hours in assisting this book's production. Bill himself is a living "history book" of the Reading Fire Department, having spent much of his lifetime with the department. Chief Rehr best exemplifies the spirit and honor of this fire department.

Coauthors Michael Glore and Michael Kitsock have arranged for the royalties from the sale of *Reading Firefighting* to benefit the Reading Fire Museum. The Reading Fire Museum is located at the Liberty Fire Company No. 5 station in southwest Reading. This magnificent Victorian station house contains and displays some of the richest historical firefighting artifacts in the United States. The royalties will contribute to the restoration and refurbishment of this important architectural and historical edifice.

FOREWORD

It is my privilege to serve as the 15th fire chief of the City of Reading, Pennsylvania, since the position was first created in 1867. It is equally gratifying to provide the archival information for the photographs on the following pages, which so richly illustrate the department's 234-year history. Looking at the work of my predecessors and their transitional accomplishments over the past years, I find that leading this department today in the 21st century is an equally daunting task. The volume of calls has multiplied by seven times just since my entry into the department as a firefighter in 1959. Unfortunately, this is typical of many older cities in the eastern United States, as the demographics of these areas change.

At one time Reading was one of the largest cities in the country protected by a predominantly volunteer fire department, using paid drivers to operate the apparatus. Today it is the opposite, with the paid career force having grown to 150 personnel and just a handful of certified volunteers. The 14 volunteer fire companies, which originally had 14 individual fire stations with 30 apparatus and ambulances, have been "right-sized" by city government to eight stations housing 11 in-service apparatus and three paramedic ambulances. Many companies have been merged into joint facilities, and some no longer have an apparatus of their own. New trucks being purchased no longer have fire company names on them. The volunteer companies, which once boasted of membership rolls totaling 10,000 people, 200 of which were "active" firemen, have shrunk drastically, with only 32 volunteers certified to fight fires. Few companies are able to assemble a quorum to hold a monthly business meeting.

Nonetheless, the volunteers and paid personnel who served this department over its long history prevented Reading from suffering a major conflagration, which befell many cities, particularly during the 19th century. Much credit must be given to those dedicated firefighters who worked with far less sophisticated apparatus and tools than we enjoy today. It is easy to underestimate the difficulties they encountered, until one understands how the fire service operated in the past. This book is dedicated with profound gratitude to all those who have served the Reading Fire Department since 1773.

—William H. Rehr III
Fire Chief, 2007

The first fire engine in Reading was this hand-drawn apparatus manufactured by Patrick Lyons in Philadelphia, and delivered to the Rainbow Fire Company in 1820. It is currently on display at the Historical Society of Berks County in Reading, at Centre Avenue and Spring Street.

One
Reading Towne Organizes a Fire Department

During the Colonial period of American history, a small settlement along the Schuylkill River in Pennsylvania became known as Reading Towne. With its important location and proximity to Philadelphia, Reading Towne prospered. Fearful of the devastating effects of fire on their developing community, the citizens of Reading Towne organized the Union and Rainbow companies in 1771 and 1773, respectively. Thus, less than 40 years after Benjamin Franklin organized the nation's first volunteer fire company in nearby Philadelphia, the Reading Fire Department emerged.

During the early 1800s, several new volunteer fire companies formed and soon disbanded. These companies included the Sun, Lafayette, and Protection Fire Companies. Sons of Reading's oldest surviving company, the Rainbow, organized the Junior Fire Company in 1813. Early hose companies included the Union Hose, Hope Hose, Ringgold Hose, and Reading Hose Company, the sole survivor of the early hose companies. With Reading's population swelling, additional companies formed: the Neversink No. 3 in 1829, the Friendship No. 4 in 1848, and the Liberty No. 5 in 1854. In 1855, the Washington Hose Company No. 2 and the Keystone hook and ladder company organized.

The advent of steam power brought remarkable changes to the Reading Fire Department, commencing with the arrival of Reading's first steam fire engine in 1860. The Reading Hose Company No. 1 purchased a steamer built in New York City for $3,000 in October 1860. The Rainbow Fire Company No. 1 followed with two steamer acquisitions in 1863, a locally built steamer and an Amoskeag engine. The Amoskeag Company sold many early steam fire engines to the Reading Fire Department. The Juniors added an Amoskeag steamer to their equipment roster in 1864; the Liberty and the Friendship companies acquired Amoskeags in 1865 and 1867.

New companies organized during the latter 19th century: the Hampden No. 6 in 1867, the Marion No. 10 in 1881, the Riverside No. 11 in 1891, the Schuylkill No. 12 in 1892, and the Union No. 13 in 1898. Reading's youngest fire company, the Oakbrook No. 14, organized in 1902. Reading's earliest steamers were originally hand-pulled to the fires. Hand-pulling fire apparatus required a Herculean effort by company members. Horses soon replaced the manpower for moving apparatus. The first company to use horses in responding to a fire alarm was the Junior Fire Company No. 2. Soon the other companies added horses for pulling not only the heavy steamers, but also hose and chemical wagons.

The Reading Hardware Company complex at Sixth and Canal Streets lies in ruin after the biggest fire in Reading's history ravaged the business on July 2, 1888. The glow was seen in the nighttime sky as far away as Lancaster and Pottsville. Juveniles threw fireworks through an open window, causing the fire. The complex was rebuilt and fully operational within a year. This view was taken from the 500 block of Canal Street the day after the fire.

The Reading Hose Company's 1875 Silsby steam engine pumps water from the fire hydrant in the 600 block of South Sixth Street, near Canal Street, the morning after a fire ravaged the Reading Hardware complex the night of July 2, 1888.

A gutted shell is all that remains of the Nolde and Horst Company hosiery mill, 820 Moss Street, after a general-alarm fire on December 7, 1899. One employee died and numerous others were rescued by ladders. The building was rebuilt and expanded and is now part of the block-square Moss Street outlet building.

George Washington "Bob" Miller served as Reading's fire chief from 1886 until 1916. Miller was a member of the Keystone Fire Company whose tenure included the transition from horse-drawn to motorized apparatus and Reading's biggest fire, which occurred at the Reading Hardware Company on July 2, 1888.

The Schuylkill Fire Company's horse-drawn 1899 Holloway chemical combination is shown in front of its station, constructed in 1898 on the southwest corner of Schuylkill Avenue and West Green Street.

The Schuylkill Fire Company's 1894 first-class Ahrens steamer, with its two-horse hitch poses in front of the station at the southwest corner of Schuylkill Avenue and West Green Street. The steamer was replaced in 1921 by a motorized American LaFrance 750-gallon pumper.

The Union Fire Company's 1911 Holloway chemical combination wagon participates in a parade on Penn Square in downtown Reading in August 1917. The driver is identified as William Haas.

The Union Fire Company's 1909 American LaFrance horse-drawn steamer participates in a parade in the 400 block of Penn Square around 1915.

Members of the Liberty Fire Company built the first two floors of the current fire station themselves in 1876, when the city did not have the money to replace its first station that was built in 1854. The company added a third floor in 1895 and removed the bell tower. The building is still an operating fire station today, located on the southeast corner of Fifth and Laurel Streets. It is also the home of the Reading Area Firefighters Museum.

The Keystone Fire Company's 1909 horse-drawn American LaFrance 75-foot hook and ladder is raised to the sixth floor of the Wertz Cold Storage building at Front and Franklin Streets, shortly after the apparatus was purchased. A motorized LaFrance tractor was added to the truck in 1917 to replace the horses.

The Keystone Fire Company's new 1909 American LaFrance hook and ladder sits in the 100 block of Franklin Street shortly after being purchased. The three-horse hitch was replaced by a LaFrance motorized tractor in 1917.

The Gamewell electric fire alarm system was installed and fully operational on June 3, 1873. Visual indicators with gongs, like the one shown, were installed in all fire stations. The cost of one of these electric marvels (before public electricity and telephones existed) was $1,250. At that time a row home in the city could be purchased for $850. Box 236 was at Eighteenth Street and Mineral Spring Road in East Reading.

Connected to the Gamewell fire alarm system were 10 tower bells throughout the city, some on fire stations, like this one at the Washington station on Spruce Street, and some in church towers. The number of the fire alarm box would be struck on the bells four times alerting the public as well as volunteer firemen.

The Liberty Fire Company's 1881 Silsby steamer and its 1897 chemical combination sit in readiness inside the engine room, with the "quick hitches" for the horses suspended from the ceiling, around 1900. The metal track in the floor under the steamer wheels assured that the apparatus would track straight through the narrow bay door upon exiting.

The Liberty Fire Company operated a street sprinkler, beginning in 1872. It was used to sprinkle the dirt streets to settle the dust and wash dirt to the gutters. The weekly fee was 5¢ per property, and those who refused to pay were blacklisted to the extent that the driver shut off the sprinkler, letting dust and dirt lie, much to the chagrin of the neighbors.

The Washington Fire Company's 1886 three-horse Hayes 65-foot hook and ladder and its 1897 Holloway chemical wagon sit in front of the company's station at 1019 Spruce Street, which was built and occupied in 1880. After remodeling the building twice, it was finally vacated by the city on December 3, 2004.

Only some of the walls remain of the Mohn Brothers Hat Factory, at 213 South Eleventh Street, after a raging fire broke out in the evening of February 13, 1899, at the height of a blizzard. Several firemen were hurt when a large pile of snow and slush fell from the roof of the building during the fire.

The Liberty Fire Company's 1881 Silsby steam engine undergoes a pumping test at Jason's Locks, on the Schuylkill Canal, near Sixth and Canal Streets around 1885. The Washington Fire Company's 1875 Leverick horse-drawn hook and ladder is parked to the right.

The Schuylkill Fire Company's 1894 Ahrens steamer sits in front of the company's first building (made of iron), which was constructed on the southwest corner of Schuylkill Avenue and West Green Street in 1894. The building was replaced by a new one in 1898, which still stands today but is now occupied by a plumbing contractor. The Schuylkills merged with the Riverside company at McKnight and Spring Streets in 1976.

The Schuylkill Fire Company's 1899 Holloway chemical combination apparatus participates in a parade in the 400 block of Penn Square in 1915.

Assistant Chief Charles Stoner, a member of the Neversink Fire Company, who served under Fire Chief John Neithammer from 1920 to 1924, poses with his horse and buggy at an unknown location. Stoner later went on to become a city councilman and director of public safety.

The Washington Fire Company's 1886 Hayes ladder truck is seen heading east on Spruce Street near South Eleventh Street at a full gallop. This truck replaced its first ladder truck, which was an 1875 Leverick, made in Brooklyn, New York.

The Reading Hose Fire Company, established in 1819, began the department's ambulance service in November 1887, with this one-horse ambulance made locally by G. W. Biehl, shown here in front of the station in the 600 block of Franklin Street.

Neighbors line the 100 block of Exeter Street as the Riverside Fire Company's 1904 American LaFrance horse-drawn chemical combination proceeds down the street, apparently headed for a parade in the early 1900s. The company's station at Front and Exeter Streets is the tall building in the background.

John G. Neithammer served as the city's fire chief from 1916 to 1932. He was a popular gentleman who ran an upscale saloon on North Eighth Street near his home fire company, the Rainbows. He was best known in the department for his pioneering efforts in fire prevention, perhaps because during his tenure in the Roaring Twenties, the city experienced a rash of costly, spectacular fires.

The Rainbow Fire Company No. 1 was organized on March 17, 1773, making it one of the oldest volunteer fire companies in the country. Pictured here in front of their station at Eighth and Court Streets, built in 1870, is the company's 1906 Nott steamer and an 1894 hose cart. The steamer was fitted with a two-wheel, four-cylinder LaFrance motorized tractor in 1917. The fire station was replaced in 1950 with a new building on the same site, which is still in use today.

The first horse-drawn chemical engine (Reo) in the Reading Fire Department was acquired by the Keystone Fire Company in 1888, pictured here in front of its station on the southeast corner of Second and Penn Streets. The apparatus's ability to go into service immediately upon arriving at a fire was its distinct advantage over the steam engines, which required time to build up steam pressure.

Fire Chief John Neithammer poses with members of the Oakbrook Fire Company in the 600 block of Park Avenue on September 14, 1923, the day on which the company became motorized with a 1923 American LaFrance 750-gallon pumper. Shown here is the "Oakies" horse-drawn chemical combination wagon that originally belonged to the Riverside Fire Company.

The Washington Fire Company's second hook and ladder was this three-horse-hitch truck made by the Hayes Company in 1886. The truck was eventually replaced by a motorized LaFrance ladder in 1917.

The Junior Fire Company's 1880 first-class Clapp and Jones horse-drawn steam engine stands in front of the company's station at Vine and Walnut Streets, built in 1877. A new station was erected at the same site in 1939, as a WPA construction project. Today it is the department's Emergency Medical Service (EMS) station, housing all of the city's ambulances.

The Neversink Fire Company's 1887 American LaFrance horse-drawn steam engine is pictured in front of the company's station, built in 1884 on the northeast corner of Third and Court Streets.

Snow covers the figure on top of the Firemen's Monument, erected in City Park in September 1901. Legend has it that the fire chief at the time, George Washington "Bob" Miller, posed for the figure and its image is his likeness.

Inside the Junior Fire Company's engine house, at 638 Walnut Street, one of the horses assigned to the company's 1893 Clapp and Jones steamer casts a wary glance at a motorized 1912 Knox chemical combination engine. All of the city's horse-drawn apparatus were replaced with motorized vehicles between 1911 and 1923.

The Union Fire Company's 1923 American LaFrance 750-gallon pumper, driven by Stanley Bechtel, precedes the company's horse-drawn 1909 LaFrance Metropolitan steamer that it replaced in September 1923. The company's uniform marching unit follows the apparatus on Penn Square in a Labor Day parade in 1924.

Two

THE DEPARTMENT MOTORIZATION

The advent of the 20th century brought more profound changes to the Reading Fire Department. As the internal combustion engine improved in reliability and horsepower, fire companies gradually replaced their horses with gasoline-powered apparatus. Several Reading fire companies began using motorized equipment in 1911 and 1912. The Liberty No. 5 purchased a motorized American LaFrance chemical fire engine in 1911, and the Junior No. 2 added a Knox chemical/hose truck in 1912. Also in 1912, the Neversink No. 3 added a Christie motorized tractor to its American LaFrance steamer, and the Schuylkill Fire Company No. 12 placed a motorized chemical truck in service. The Reading Hose Company No. 1 also placed into service a Knox chemical/hose truck late in 1912.

With the population of the city of Reading peaking over 100,000 in the early 20th century, rapid motorization of the department continued. In 1913, the Friendship No. 4 purchased a White chemical/hose truck, and the Hampden No. 6 added an American LaFrance engine capable of pumping 1,400 gallons per minute, the department's largest pumper. The Keystone hook and ladder company obtained an American LaFrance tractor for its ladder truck in 1916, and the Rainbow No. 1 purchased an American LaFrance tractor for its Nott steamer in 1917. Also in 1917, the Washington No. 2 housed a new motorized ladder truck, and in 1918, the Marion No. 10 purchased an American LaFrance engine.

The horse-drawn era for the Reading Fire Department ended in 1923. The final companies to motorize were Riverside No. 11, Union No. 13, and the Oakbrook No. 14. The Riverside placed an American LaFrance in service in 1921, and the Union and Oakbrook companies both motorized in 1923. The Union No. 13 holds the distinction of being the last company in the department to use horses. On September 15, 1923, the Union placed into service an American LaFrance triple-combination pumper. By October 11, 1923, the last three Reading fire horses, Joe, Dick, and Tom, were reported sold and the horse stalls removed.

Several important Reading businesses also organized private fire brigades during the early 20th century. The largest private fire brigade was the Reading Railroad Fire Department. This fire brigade consisted of many city volunteer firefighters who were employed by the Reading Railroad Company. The brigade provided fire protection to the vast Reading Railroad shops and yards located in the northern part of the city.

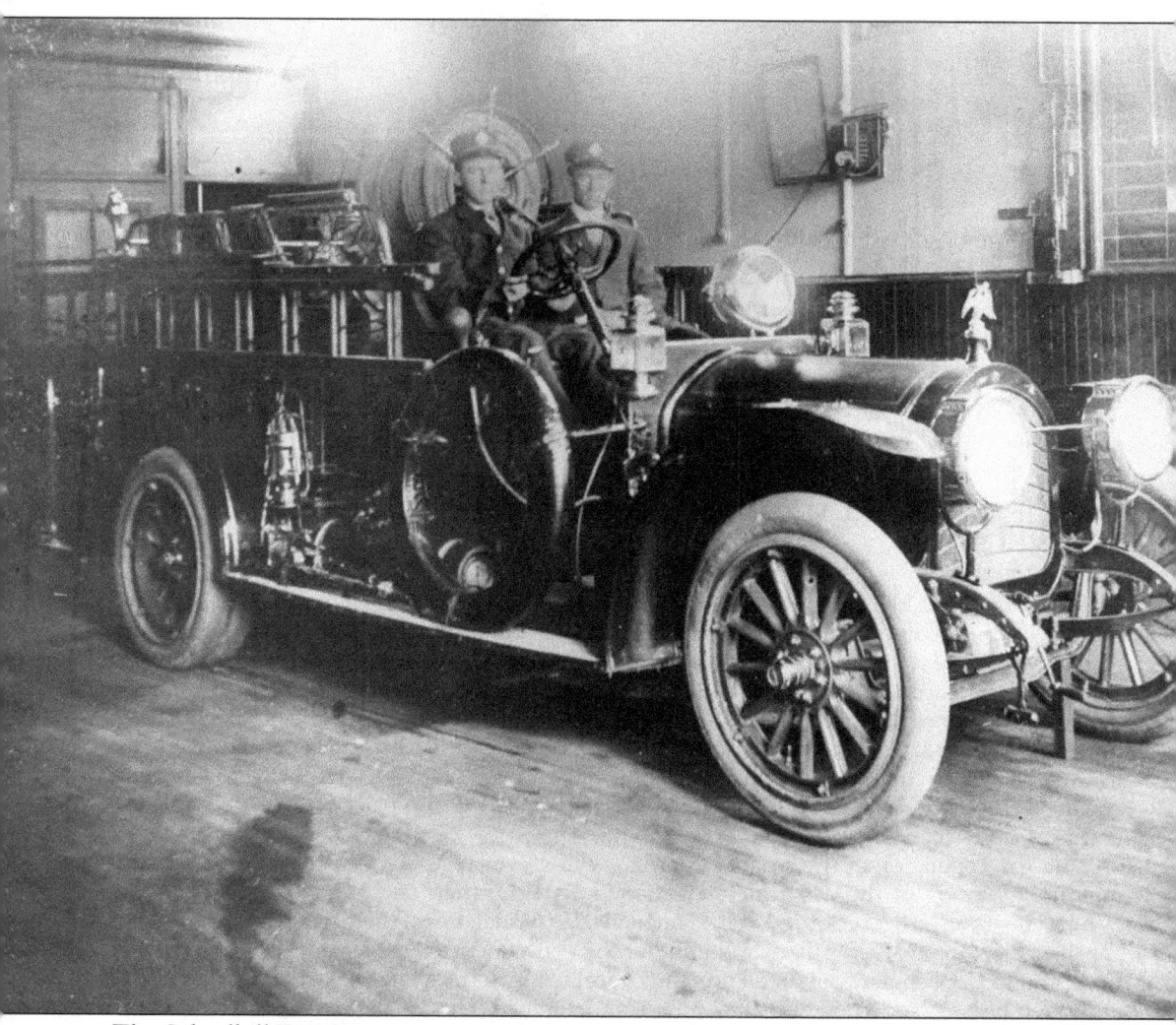
The Schuylkill Fire Company, established in 1892 in the city's northwest section, received this motorized Rambler chemical combination truck in January 1912. At the time, the company also operated an 1894 horse-drawn Ahrens steam engine from its Schuylkill Avenue and West Green Street station. A new joint station with the Riverside Fire Company was eventually built in 1976, at McKnight and West Spring Streets.

Members and apparatus from the Rainbow Fire Company pose in front of their station at Eighth and Court Streets around 1920. The horse-drawn 1906 Nott steamer was motorized with a 1917 LaFrance tractor, and the hose carriage was replaced in 1918 with this 350-gallon-per-minute White pumper.

The Keystone Fire Company's 1935 Ahrens-Fox 85-foot ladder reaches to the roof, while firemen from eight companies battle this smoky two-alarm fire at the Brighter Furniture Store, located on the southeast corner of Third and Penn Streets on November 28, 1939, at 8:00 in the morning. In the right foreground is the Neversink Fire Company's 1918 White 500-gallon pumper, while on the left the Neversink's 1925 American LaFrance 750-gallon pumper and the Reading Hose's 1924 chemical combination are nose to nose.

The Washington Fire Company, established in 1855, displays its 1917 American LaFrance motorized tiller ladder and its 1919 LaFrance/Holloway chemical engine at the company's station, built in 1880 at 1019 Spruce Street.

The Central and Belmont Hotels, on the northeast corner of Fifth and Court Streets, burn during the night of March 11, 1929. One person died while others jumped into life nets. The downtown branch of the Reading Post Office now occupies this site.

The Junior Fire Company's 1912 Knox chemical combination was housed on October 12, 1912, and served until 1924, when it was declared unfit for service. It was replaced by a 1925 American LaFrance 750-gallon pumper.

The Junior Fire Company received this Seagrave 750-gallon pumper in 1928. It replaced the company's 1897 Clapp and Jones steamer that had been fitted with a motorized tractor.

The Liberty Fire Company's 1937 Pirsch 500-gallon pumper is shown inside the company-owned engine house at Fifth and Laurel Streets. The historic station, built in 1876, is now the site of the Reading Area Firefighters Museum.

White-colored apparatus was the trademark of the Liberty Fire Company from its first motorized engine in 1911 until the 1960s, when the department's white-over-red combination became the standard color scheme. Pictured here on the left is the company's 1931 Buffalo 1,000-gallon pumper and on the right is its 1937 Peter Pirsch 500-gallon pumper.

An employee stands amid the wreckage of Wire Mill No. 1, of the Carpenter Steel Company, west of Front and Exeter Streets, following a general-alarm fire in the early morning of October 6, 1928. The fire, which followed an explosion, destroyed the 100-foot-by-700-foot wire mill and annealing departments.

The Washington Fire Company's new 1930 Mack chemical combination apparatus is displayed in front of city hall at 815 Washington Street. The firefighter on the back step is Francis G. Drexler, who eventually rose to the ranks of assistant chief (1948), deputy chief (1954), and finally the city's first fire marshal (1965).

This 1913 American LaFrance 1,400-gallon pumper was only in service for 11 years, as its pumping capacity proved too powerful for the water distribution system in northeast Reading. It was replaced by a 750-gallon LaFrance in 1925.

Heavy smoke blankets the Brighter Furniture store on the southeast corner of Third and Penn Streets on November 28, 1939, shortly after 8:00 a.m. The Washington Fire Company's 100-foot aerial ladder extends to the roof of the adjoining building on Penn Street, while two streams from two-and-a-half-inch hoses are directed into the third-floor windows.

The Riverside Fire Company, established in North Reading in 1890, displays its 1921 American LaFrance 750-gallon pumper at the company's station, at the northeast corner of Front and Exeter Streets, built in 1896. The company merged with the Schuylkill Fire Company in a new station at McKnight and Spring Streets in 1976.

Fire driver Paul Mogel, patriarch of the Mogel firefighting family, is shown at the wheel of the Riverside Fire Company's 1921 American LaFrance 750-gallon pumper in March 1933. Mogel drove at the Riverside station at Front and Exeter Streets for 40 years.

The Marion Fire Company, established in 1881, became the city's third ladder company in 1915 with the acquisition of this motorized 75-foot Boyd tiller-type ladder truck. In 1921, an American LaFrance tractor replaced the Boyd front end. The company's horse-drawn American LaFrance steamer and Holloway chemical, both purchased in 1889, are shown in the background in front of the station at Ninth and Marion Streets.

The Union Fire Company, established in East Reading in 1898, is shown at its station at Fifteenth and Muhlenberg Streets with the company's 1923 American LaFrance 750-gallon pumper. The Unions were the last to use horses on September 15, 1923, when their 1909 horse-drawn LaFrance Metropolitan steamer was replaced by the apparatus shown above.

Fire Chief John Neithammer, wearing the white "Chief F. Dept." coat directs firefighters in wetting down the rear of the Wertz Milling Company, 137–139 West Buttonwood Street, after a general-alarm fire in the afternoon of June 7, 1920. This view is looking west on Schiller Street toward Division Street. The fire also damaged several adjoining buildings and nearby railroad cars on a siding.

The Reading Hose Fire Company's c. 1920 apparatus is shown in front of its Franklin Street station. On the left is its 1918 Studebaker ambulance, in the middle is a 1912 Knox chemical combination, and on the right is its 1892 Silsby steam engine with a 1913 Knox motorized tractor replacement for the horses.

The Oakbrook Fire Company was originally organized as the Liberty Fire Company of Milmont in Cumru Township in 1902, before it was annexed to Reading in 1914. After annexation, the company changed its name to Oakbrook. The company was motorized one day before the Union Fire Company, on September 14, 1923, with the American LaFrance 750-gallon pumper shown.

The Schuylkill Fire Company's fully motorized apparatus is shown in front of its decorated fire station at Schuylkill Avenue and West Green Street. On the left is the company's 1921 American LaFrance 750-gallon pumper, and on the right is its 1912 Rambler chemical combination.

Reading's first high-rise building fire occurred on February 10, 1930, when fire broke out on the roof of the newly built Abraham Lincoln Hotel at Fifth and Washington Streets. Looking worse than it was, the fire involved hot pitch and roofing material on the roof, which was quickly brought under control in 40 minutes.

The Friendship Fire Company's 1923 American LaFrance 750-gallon pumper (left) and its 1913 White chemical combination (right) are shown in front of the company's station, built in 1873, on the northwest corner of Peach and Franklin Streets. In 1931, a new station was built on this site, which was razed in 1972 during urban renewal. The company was housed temporarily with the Rainbows until it was merged with the Reading Hose Company in a new station, at Plum and Franklin Streets in 1975.

Although no damage is apparent from the front, a fire resulting from an exploding painter's blowtorch caused substantial damage to a second-floor apartment kitchen at 48 North Fifth Street on March 17, 1948. In the foreground is the Washington Fire Company's 100-foot Mack/Pirsch tractor-drawn ladder, and to the rear of it is the Keystone Fire Company's 1944 Mack/LaFrance hybrid engine. Fire Chief Edward C. Dell and Assistant Chief Francis G. Drexler's coupes are nose to nose alongside the ladder.

Fire Chief Edward C. Dell leads an occupant of 48 North Fifth Street to safety after he refused to leave his second-floor apartment on March 17, 1948. Dell climbed the Washington Fire Company's aerial and escorted the victim to safety. An exploding painter's torch caused a fire that substantially damaged a second-floor kitchen.

This is the Junior Fire Company's alarm desk area as it appeared shortly after the company occupied its new building at 638 Walnut Street built by the WPA in 1939. Above the alarm response cabinet is the bottom of the secondary circuit indicator/gong ("slow time" for tower bells), while below it, a Gamewell slash register and "joker" bell on a primary circuit announce the incoming telegraph alarm in "fast time." The driver is unidentified. Slow time and fast time refer to the speed at which the telegraph signal is received.

The Junior Fire Company became the second fire company in Reading to provide ambulance service, beginning in 1908. (The Reading Hose was first in 1887.) Shown above is the Junior Fire Company's 1930 Cadillac ambulance in front of the Reading Hose engine house on Franklin Street.

Firefighters standing atop the ice-encrusted Keystone Fire Company's 1935 Ahrens-Fox ladder truck man two-and-a-half-inch hose lines while the truck's 85-foot wooden ladder extends to the roof of the Fairmore Music Building at 135 South Fifth Street. The general-alarm fire occurred at the height of a raging snow-and-ice storm on January 28, 1948, at 7:00 in the evening.

The Keystone Fire Company's 1935 Ahrens-Fox 85-foot hook and ladder is profiled on South Second Street, below Penn Street, alongside the fire station built in 1887. This apparatus was replaced in 1959 by an 85-foot Mack/Maxim tractor-drawn tiller ladder.

The Washington Fire Company's 1935 Mack aerial ladder (left) and the Keystone's 1935 Ahrens-Fox aerial (right) operate at a general-alarm fire at the Mohican supermarket at 718–720 Penn Street, on April 18, 1947.

The Neversink Fire Company's 1925 American LaFrance 750-gallon pumper is shown at the company's engine house on the northeast corner of Third and Court Streets where it stood from 1886 until it was replaced by a consolidated station with the Keystone Fire Company across the street, on the southeast corner, in 1963.

In May 1940, the city purchased this Type 80 model Mack 750-gallon pumper to replace the Neversink Fire Company's 1918 White pumper. It is one of the few Type 80 model Macks built at that time. Daniel Lengel is believed to be the driver pictured.

The Keystone Fire Company's 1925 LaFrance chemical truck (left) and its 1935 Ahrens-Fox 85-foot hook and ladder are shown at the front of the engine house on the southeast corner of Second and Penn Streets. The station, completed in 1887, is still standing, now occupied by a private business. The fire company was merged into a consolidated station with the Neversink Fire Company at Third and Court Streets in 1963.

The Keystone and Neversink Fire Companies were the first to be merged into a new consolidated fire station, built on the southeast corner of Third and Court Streets in 1963. Pictured from left to right are the Neversink's 1940 Mack 750-gallon pumper, its 1959 Mack 750-gallon pumper, and the Keystone's 1959 Mack/Maxim 85-foot tractor-drawn aerial. Next to the ladder are Deputy Chiefs Merle Gerlach and Russell P. Mogel (who eventually became fire chief in 1971).

The Washington Fire Company's 1917 LaFrance ladder truck was replaced with this 100-foot 1939 Pirsch aerial ladder, powered by a 1935 Mack tractor. The tractor portion was originally attached to the old LaFrance ladder from 1935 to 1939. The view is looking east on Spruce Street with the station behind the ladder on the left.

As daylight dwindles, so does the crowd on Penn Square, attending the annual Labor Day fire department inspection in 1950. Over the years, the Labor Day inspection of the fire department also included aerial ladder maneuvers and pumping evolutions.

The only Ward LaFrance ever owned by the city was this 750-gallon pumper assigned to the Reading Hose Fire Company in 1954. It was purchased with civil defense money at the close of the Korean War.

The Reading Hose Fire Company's 1954 Ward LaFrance pumper passes under a ladder arch on Penn Square, formed by members and apparatus of the Washington and Marion Fire Companies, during a Labor Day inspection in 1960.

A stubborn, smoky one-alarm fire damaged the Craft Films store at 102–104 North Fourth Street in the early evening of October 2, 1957. Fire Chief Russell C. Bowers was overcome by smoke while fighting the fire.

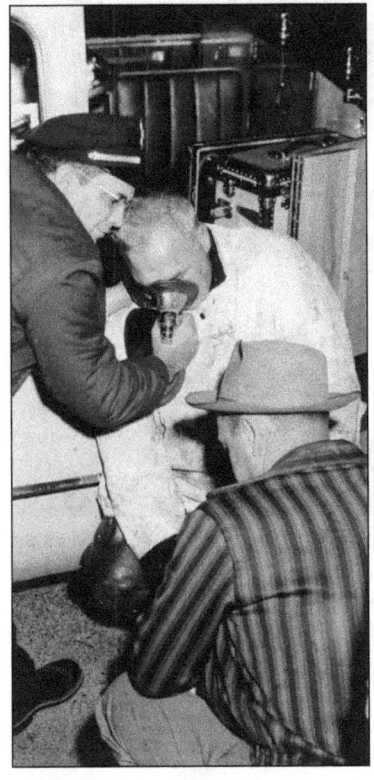

Fire Chief Russell C. Bowers receives oxygen from Schuylkill Fire Company ambulance driver Kenneth "Candy" Clouser at a stubborn, smoky fire at the Craft Films store at 102–104 North Fourth Street on October 2, 1957.

Edward C. Dell became the department's 10th fire chief on January 5, 1948, having served as an assistant chief since 1931. On December 2, 1953, Dell was stricken on the way to a call and died in the center of the city on Penn Square. A passenger riding with him managed to bring his car to a stop. Because he did not complete the call, legend has it that his ghost continues to haunt his home fire station, the Hampden Fire Company, at Eleventh and Greenwich Streets, which is now a privately owned apartment building.

Other than the 1937 Diamond-T rescue truck at the Friendship Fire Company, the Union Fire Company's 1949 Mack 750-gallon pumper was the only closed-cab apparatus operated in the department until the 1960s. The Union company paid the city the extra cost for the cab out of its "sinking fund" monies, raised from carnivals it had held over the years. The other two 1949 Macks the city bought were open cabs.

The Friendship Fire Company's 1949 Mack 750-gallon pumper passes under a ladder arch formed by the Washington and Marion Fire Company aerial ladders on Penn Square during a Labor Day inspection in 1962.

The three fire company ambulances pose in a rare group photograph behind the Community General Hospital at Sixth and Walnut Streets around 1950. From left to right are the Reading Hose, Schuylkill, and Junior units. The three companies provided ambulance service until 1974, at which time the city took over full operation of EMS and established one of the region's first paramedic services.

A Metropolitan Edison Company lineman watches as a general-alarm fire rages through the four-story factory-mill construction M. J. Earl warehouse, at Seventh and Walnut Streets on November 29, 1955. Next to the lineman is pedestal-mounted Gamewell fire alarm box No. 36, which was pulled for the fire at 3:15 p.m.

The Friendship Fire Company, established in 1848, became a specialty rescue company in 1938 with the arrival of a Diamond-T "emergency squad" truck, shown above. The company also continued to operate a 1923 LaFrance engine and later a 1949 Mack 750-gallon pumper. The city's "fifth generation" heavy rescue replacement vehicle is due for delivery in the late summer of 2007.

The Friendship Fire Company's second rescue truck, replacing the original 1937 Diamond-T, was this 1957 GMC unit, which became affectionately known as the "covered wagon," shown here at a Penn Square Labor Day inspection. It ran for 19 years until it was replaced by a 1976 Hamerly rescue truck.

Two children, aged 2 months and 18 months, died during this mid-day fire at 22 Neversink Street on March 31, 1950. Pictured in the foreground is the Friendship's 1949 Mack and the Reading Hose's 1928 LaFrance, and in the back from left to right are the Keystone's 1944 Mack/LaFrance engine, the Liberty's 1937 Pirsch, and the Keystone's 1935 Ahrens-Fox ladder.

Lightning from a passing summer thunderstorm sparked this general-alarm fire that destroyed the Sam Strause Lumber Company at 416 Blair Avenue in the city's Glenside section on August 21, 1958.

Heat to many downtown buildings was interrupted on this cold February 12, 1958, morning when fire heavily damaged the Reading Steam Heat Company in the 600 block of Elm Street. The Keystone Fire Company's 1944 Mack/LaFrance hybrid engine can be seen in the foreground, while other members scale the company's 1935 Ahrens-Fox wooden aerial in the background.

The Junior Fire Company, established in 1813, received this 750-gallon B-85 Mack in 1959, shown here across from its station at Reed and Walnut Streets.

Russell P. Mogel, foreman of the Riverside Fire Company (to the right of the wreckage, with white coat and helmet on) surveys the damage of one of two fighter planes that collided in mid-air and crashed in Gethsemane Cemetery on December 11, 1948. Both pilots were killed. The Reading Fire Department responded as a mutual-aid department with Muhlenberg Township companies. Mogel eventually became the city's 12th fire chief in 1971.

The Keystone Fire Company's pumper was a Mack/LaFrance hybrid, which was the result of an accident. It was originally purchased as a 1925 LaFrance chemical truck. In 1940, the chemical tanks were removed and a pump and booster tank added. In June 1944, it was involved in a serious accident at Fifth and Bingaman Streets, which destroyed the LaFrance chassis. The Mack Corporation was contracted to provide a new chassis, retaining the LaFrance pump and booster tank.

Fire driver William "Socks" Becker (left) and an unidentified driver inspect the equipment on the Reading Hose's 1937 Sayers and Scovill ambulance.

In August 1937, the Reading Hose Company received this Sayers and Scovill ambulance, equipped with air conditioning, heating, a radio, and a respirator. The driver is unidentified.

This was the scene that greeted the first-arriving companies at the Reber Potato Chip Company at 221 Church Street in the early-morning hours of March 14, 1956. The entire building was destroyed in the general-alarm fire.

The Keystone Fire Company's 24-year-old Ahrens-Fox 85-foot wooden aerial ladder truck was replaced with this 1959 Mack/Maxim 85-foot steel ladder. It operated from the Third and Court Streets station for 25 years, until it was replaced by a 1983 Seagrave rear-mount 100-foot aerial.

The Glidden Paint Company warehouse in the rear of the company's plant at Third and Bern Streets is heavily involved on this hot summer night on July 28, 1954. The fire was set by a member of the Riverside Fire Company, who was later arrested and sent to prison.

The Liberty's 1937 Pirsch engine and the Keystone's 1935 Ahrens-Fox operate at a two-alarm fire, which heavily damaged the Belle Chemical Company complex in the 500 block of Pearl Street on January 8, 1959. The building in the foreground survived, only to be completely destroyed in an early-morning fire on November 21, 1967.

The Schuylkill Fire Company became the fourth ladder company in the department with the acquisition of this 75-foot Peter Pirsch aerial. In April 1956, fire drivers Joseph Gallagher and Raymond Beisswanger drove the apparatus from the Pirsch plant in Kenosha, Wisconsin, to Reading.

Firefighters using the Schuylkill Fire Company's 75-foot Pirsch aerial continue the overhaul and mop-up at the Bloom Furniture warehouse at 128 Wood Street on June 11, 1957. The general-alarm fire started at a truck loading dock adjacent to the building and spread to the three-story warehouse.

Three
MODERN ERA

As the 20th century progressed, the Reading Fire Department faced new challenges. An eroding industrial tax base, changing demographics, and dwindling ranks of volunteer firefighters posed new concerns for city and fire department leaders. Additionally, Reading Fire Department fought several major blazes in the 1960s and 1970s as abandoned industrial sites became easy targets for arsonists. The city responded to these challenges with larger, more diverse apparatus, company mergers, and eventually a paid department. The era of the gasoline truck engine virtually ended with the diesel engine becoming the prime power source for fire apparatus.

From the late 19th century until 1982, Reading's Gamewell fire alarm system provided a means of transmitting fire alarms within the city. The system consisted of 32 operating circuits and 196 street boxes throughout the city. Originally 10 tower bells tolled the incoming alarms as a means of alerting volunteer firefighters. House bells, indicators, and tape registers also recorded alarms at each station. The Reading City Hall alarm room received and transmitted alarms to respective stations. In 1963, the Sentinel Company took over dispatching duties. Berks County has provided dispatching at its 911 center in the Berks County Courthouse since 1990.

Beginning in 1963, fire company mergers became prevalent throughout the department. Fewer volunteers, less tax dollars, and faster, more powerful fire apparatus made mergers increasingly viable for the city. The first companies to merge were the Keystone and Neversink companies in 1963. The Reading Hose and Friendship companies united in 1974. The Schuylkill and Riverside companies joined in 1976. Four more Reading fire companies combined in 1983: the Hampden and Marion companies and the Rainbow and Junior companies. In 2002, the Washington and Union companies consolidated. The final merger took place in 2005 as the Rainbow, Junior, Washington, and Union companies formed one fire company.

As the ranks of the volunteer firefighting cadre slowly depleted, a three-platoon, predominantly paid department emerged in 1969. In this year, the firefighters' union, International Association of Firefighters Local 1803, also was organized. Volunteer firefighters would now supplement paid firefighters on emergency calls. Diverse apparatus purchased during this era included three interesting vehicles. In 1957, a Jeep brush-fire unit was placed in service with the Union Fire Company. In 1973, the Reading Fire Department obtained a Land Amphibious Reconnaissance Craft for water rescue. In 1987, a used Ford truck was converted into a hazardous materials vehicle for hazmat operations.

The second fire at the Harold's Furniture warehouse store, at Eighth and Green Streets, on January 3, 1981, completely destroyed the building. An earlier fire, caused by lightning striking the building on June 22, 1979, heavily damaged the rear portion.

A full first-alarm assignment of apparatus deploy in the 500 block of Penn Square to fight a smoky fire at the Crystal Restaurant on June 10, 1970. The fire resulted from an electrician who caused an electrical dead-short at the main breaker panel in the basement. Eleven years later, the Crystal was completely destroyed in an early-morning fire on February 26, 1981.

After the Friendship Fire Company's station at Peach and Franklin Streets was torn down in 1972 during downtown urban renewal, the company's apparatus, including the 1957 GMC rescue truck, shared quarters with the Rainbow Company's 1963 Hahn/Diamond-T 1,000-gallon pumper at Eighth and Court Streets. The Friendships eventually moved into a new consolidated station with the Reading Hose Company at Plum and Franklin Streets in 1974.

The Rainbow Fire Company operated this 1,000-gallon Hahn/Diamond-T pumper from 1963 until 1983. On January 12, 1983, it was completely demolished when it collided with the Snorkel at Eleventh and Elm Streets while responding to an alarm at St. Joseph's Hospital. The driver, Robert Krick, sustained job-ending injuries. The Snorkel was repaired and restored to service.

The fire department's officer staff in December 1963 consisted of four chiefs: Fire Chief Russell C. Bowers (seated) and, from left to right, Deputy Chief Francis G. Drexler and Assistant Chiefs Albert P. Batastini and Russell P. Mogel.

The Hampden Fire Company marching unit passes under the arched ladders of the Washington (left) and Marion (right) Fire Companies' aerial trucks during the annual fire department Labor Day inspection and parade on Penn Square on September 6, 1971.

Three occupants of Bert and Alfredo's Hotel, on the northeast corner of Third and Franklin Streets, perished during this Friday-night fire on January 7, 1972. The Washington Fire Company's 100-foot Pirsch/International ladder pipe stream douses the burning cupola on the roof, while the Schuylkill's 1956 Pirsch 75-foot ladder ventilates the third floor. The Rainbow's 1963 Hahn/Diamond-T 1,000-gallon pumper supplies numerous hose lines in front of the building.

Assistant Fire Chief Harry Stauffer (behind the tow chain) directs rescue operations to free the driver of a fuel-oil truck who lost his brakes while descending the hill on Douglass Street leading to North Eighth Street on May 7, 1969. The truck struck the building opposite the end of Douglass Street, trapping the driver in the cab.

The Keystone Fire Company's 1959 Mack/Maxim 85-foot tractor-drawn aerial operates a ladder pipe at a two-alarm fire at Garman Brother's Construction Company at 100 Neversink Street on January 10, 1968. The ladder is shown working the rear of the building on Neversink Alley. The fire occurred the day after a gas explosion and fire killed nine members of two families on North Carroll Street in the city's Oakbrook section.

A two-alarm fire, caused by a malfunctioning road grader inside the building, destroyed the N. H. Garman Construction Company at 100 Neversink Street on January 10, 1968. Ladder pipe streams from the Washington's 100-foot Pirsch aerial and the Schuylkill's 75-foot Pirsch truck work the front of the building. The Liberty's 1937 Pirsch engine and the Washington's 1951 Mack engine share side-by-side positions in front of the building.

A huge real Christmas tree in the main lobby of the Berks County Courthouse ignited and burned, apparently from an electric short in a string of lights, on December 16, 1966. The large ornamental globe of the world mounted on the ceiling was severely scorched, but the fire-resistive construction of the lobby area prevented the fire from spreading.

Deputy Chief Merle A. Gerlach, covered with foam, emerges from a flammable liquids fire in the basement of the Flying Eagles hangar at the Reading Airport on February 22, 1971. At the time, the Reading Fire Department provided primary response and protection for the Reading Airport.

Firefighters operate on North Mill Street, at the rear of the Boscov's department store fire at Ninth and Pike Streets, on February 2, 1967. The hose crew on the right appears to be operating next to an old outhouse in the backyard of an adjacent dwelling. The Boscov general alarm was the third multiple-alarm fire within five days. The Heller Bindery burned on January 29, and the Colony Inn on January 30.

The burning Boscov's department store at Ninth and Pike Streets on February 2, 1967, is shown in this spectacular aerial view taken by a *Reading Eagle* newspaper photographer in an airplane. Another Boscov store known as Boscov's West, in the western borough of Sinking Spring, burned to the ground nine months later in November.

Pomeroy's Department Store's outside garden center on the northeast corner of Sixth and Cherry Streets was destroyed during a mid-morning fire on November 4, 1965. A discarded cigarette was blamed for starting the blaze. Automatic sprinklers stopped the fire from extending to the store's interior.

The Strand Theatre at Ninth and Spring Streets burns during a two-alarm fire on Sunday morning, February 23, 1970. The theater had ceased operations a short time before the fire. A McDonald's now occupies the site.

Russell P. Mogel is sworn in as fire chief by Mayor Gene Shirk in February 1971, with six-year-old son Gary assisting. Chief Mogel retired in June 1983. Gary Mogel became a deputy chief in August 2002.

The Mogel family has been noted for its years of dedication to the fire service, both in Reading and elsewhere. From left to right are Paul Mogel, who drove at the Riverside Fire Company for 40 years; his son Russell P., fire chief from 1971 to 1983; son Robert, who rose to the rank of captain in the Washington, D.C., Fire Department; and son Richard, who drove at the Neversink Fire Company until retiring to Florida.

Throngs of people crowd Penn Square on a bright sunny Labor Day in 1960 for the annual Labor Day inspection of the Reading Fire Department. The engine in the foreground with the floodlights mounted on top of the hose bed is the Union Fire Company's closed-cab 1949 Mack 750-gallon pumper. The blue-and-cream ambulance next to it is the Schuylkill Fire Company's Cadillac ambulance.

Members of the public attend one of the annual Firemen's Memorial Services, dedicated to deceased paid and volunteer firemen. The Volunteer Firemen's Memorial Bandshell in City Park was built as a WPA project and dedicated on Labor Day 1939.

Fire Chief Russell C. Bowers is overcome while directing this smoky fire at the corner of Rose and Walnut Streets on October 20, 1970. Firefighter Harry Rheinsmith (left) steadies the chief, while an ambulance attendant prepares to administer oxygen.

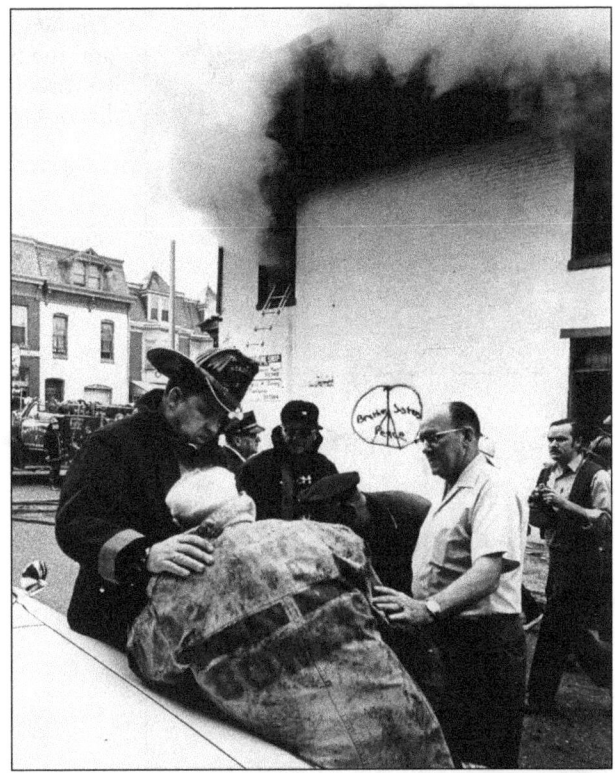

Assistant Chief Russell P. Mogel, who became fire chief in 1971, directs scores of firefighters in the rear of the Zerbe Hardware store at 649 Schuylkill Avenue on May 2, 1962. The crews shown here were operating in the 600 block of Lincoln Street and kept the fire from spreading into the three-story apartment building on the right. A large portion of the hardware store collapsed during the general-alarm fire.

The Washington Fire Company's 100-foot aerial and the Reading Hose's 75-foot Pitman Snorkel go through aerial maneuvers during the 1970 Labor Day inspection on Penn Square.

One of the new Ford sedans used by the fire chief and his assistants passes through the ladder arch on Penn Square, formed by the Washington and Marion aerial ladders, on Labor Day 1960.

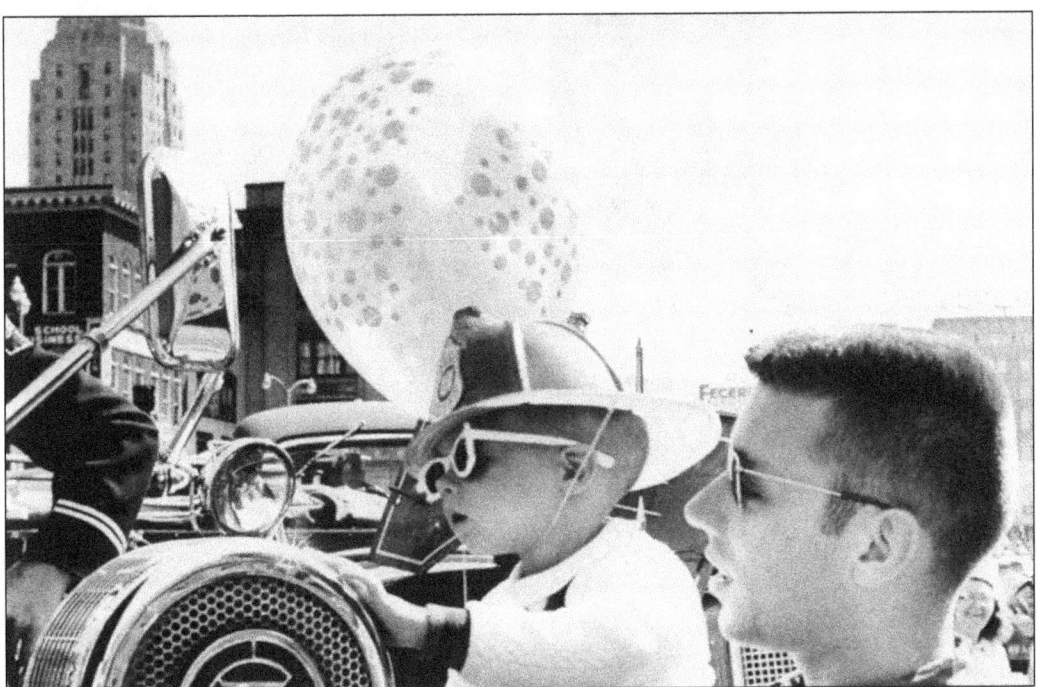

A young potential fireman inspects the Federal-Q mechanical siren on the Oakbrook Fire Company's 1957 Mack pumper during the annual Labor Day inspection of the fire department on Penn Square in 1960.

Sentinel Alarm Company, a local private security company, took over fire dispatching under contract to the city in 1963, ending 90 years of city operation. Shown here is Ronald Moyer, a Sentinel employee, checking an alarm on a newly installed console in the former city hall alarm room. One of the original Gamewell slate-board circuit testing panels still remains in the background. Sentinel eventually became part of Wells Fargo Alarm Services, which continued dispatching until the city joined the county-operated 911 center in 1990.

The Reading Hose's 75-foot Pitman Snorkel and the Washington Fire Company's 100-foot Pirsch aerial work the Fifth Street side of the O'Brien Storage Company warehouse at Fifth and Laurel Streets on October 19, 1968. During the fire, a volunteer, Anthony Miccicke Sr., collapsed and stopped breathing inside the building. He was found and revived by Deputy Fire Chief Al Batastini.

The Washington Fire Company's 100-foot Pirsch aerial (with a 1966 International tractor) deploys a ladder pipe stream on the burning Fifteenth and Perkiomen Elementary School during the early evening of December 2, 1969. The Keystone's ladder pipe can be seen on the right, operating during this general-alarm fire.

The Reading Hose Fire Company's beloved mascot, Spot, peers out of the cab of the Snorkel during one of the annual Labor Day parades.

The Reading Hose Fire Company placed this 75-foot International/Pitman Snorkel in service in April 1968. It was refurbished and upgraded over the years, including the conversion to an automatic transmission and a new KME chassis in 1989. It was finally replaced in 2003 with an American LaFrance/LTI 75-foot tower ladder. The Snorkel is still used as a reserve aerial.

Schuylkill Fire Company volunteer Lt. William H. Rehr (left) and two unidentified volunteers work a routine mattress fire with a booster line equipped with a Hardie-gun nozzle in a second-floor bedroom, in the 400 block of Gordon Street in October 1965. Rehr joined the career force in 1972 and eventually became the city's 15th fire chief in 1996.

The Schuylkill Fire Company's 1956 Pirsch 75-foot aerial provides a mutual-aid ladder pipe stream at a fire that destroyed much of the Riveredge Inn on the Bernville Road (Route 183) just north of the city line on July 30, 1971. The operator of the ladder is career firefighter Daniel Giandomenico.

Firefighters using an aerial ladder and a fire escape attack a two-alarm fire at the Colony Inn, 1635 Centre Avenue, on January 30, 1967, in the early-evening hours. The stubborn deep-seated fire in partitions and void spaces burned through much of the night. It was the second of three multiple-alarm fires that occurred within five days.

Firefighters had just been backed away by chief officers when the rear wall of the Redi-Finn Manufacturing Company collapsed during a general-alarm fire on December 8, 1969. The building was located in the rear of the 1500 block of North Ninth Street adjacent to the Reading Railroad yards.

A Neversink Fire Company unidentified volunteer makes quick work of this bread-and-butter one-story frame garage fire with a one-and-a-half-inch hand line, in the 300 block of Rose Street on September 14, 1964.

Schoolchildren from the Twelfth and Marion Elementary School get an up-close view of fire department operations at a "working" one-alarm row house fire in the 1200 block of Locust Street on April 14, 1969. A firefighter on the tip of the Marion's 1948 Pirsch 85-foot aerial ladder ventilates the third-floor dormer window. The Rainbow's 1963 Hahn/Diamond-T 1,000-gallon pumper is in the foreground.

Civil unrest, resulting from the assassination of Martin Luther King Jr. in the late 1960s, resulted in protective cages being placed on open-cab city apparatus that served "trouble areas," places where apparatus were hit with bricks and bottles. Pictured here is Schuylkill Fire Company driver Robert Spatz in the caged cab of the company's 1949 Mack pumper. The company's 1956 Pirsch aerial and the Neversink company's 1959 Mack had similar cab cages installed.

The probationary firefighter class of 1970 consisted of, from left to right, Edwin Schroeder, Edward Ott, Richard Boyer, William Buddell, and Kenneth Moyer. Boyer eventually became fire chief in 1993. All have since retired except Moyer, who currently drives the front end of Ladder 1 on the "C" platoon.

A Washington Fire Company firefighter gets ready to open the nozzle of a two-and-half-inch hose line, as the Pennsylvania Railroad freight station at Fifth and Willow Streets burns furiously during a mid-morning fire on April 20, 1968. The entire station was destroyed, but the fire did not spread to any adjacent buildings.

A ladder pipe is readied on the end of the Marion Fire Company's 1948 85-foot Peter Pirsch aerial as fire heavily involves the Weiss Brothers Scrap Warehouse at Division and Schiller Streets at 2:00 a.m. on January 5, 1972.

Three aerial trucks, the Keystone's (in the foreground with its main up), the Snorkel in front of it, and the Washington ladder in front of the Snorkel, crowd in front of a dwelling in the 700 block of Bingaman Street during a one-alarm fire on September 10, 1968. They compete for space with the Liberty's 1937 Pirsch and the Reading Hose's 1954 Ward LaFrance engines.

Sixty-mile-per-hour winds propelled a brush fire in Hampden Park, behind Reading High School, up the western slope of Mount Penn Mountain to Skyline Drive in 10 minutes. A massive response of city and county fire companies prevented the fire from jumping the road. The Washington Fire Company's 1951 Mack 750-gallon pumper (foreground) and the Hampden Fire Company's 1958 Mack 750-gallon pumper (behind it) are joined by several county companies on Skyline Drive near the Mount Penn Fire Tower.

A full first-alarm assignment works an all-hands fire in some vacant buildings near Fourth and Washington Streets on May 30, 1962. The buildings, slated for urban-renewal demolition, were the scene of several fires before they were torn down.

Fire personnel begin wrapping up after extinguishing a fire at Lobel's store in the 500 block of Penn Square on January 4, 1969. Early detection and a quick attack by arriving companies prevented the one-alarm fire from becoming another Penn Street burnout.

Less than 12 hours after the Reading Merchants Oil Company fire, another two-alarm fire struck the Raylon Beauty Supply and Bryland Beauty School buildings at 910–912 Penn Street just after midnight on January 17, 1970.

A two-alarm fire rages through the top-floor apartments above the Raylon Beauty Supply Company and the Bryland Beauty School at 910–912 Penn Street on January 17, 1970. Miraculously the tenants in the apartments all escaped unharmed.

First-arriving companies go into operation at a general-alarm fire, which ultimately destroyed the Elm and Moss Elementary School in the early-morning hours of April 30, 1960. The arsonists were later apprehended after starting another fire in the suburban Sinking Spring Elementary School.

Assistant Fire Chief Russell P. Mogel supervises the fire attack on the roof of the Pear and Company scrap warehouse at 334 Reed Street during a two-alarm fire on October 24, 1962. Mogel became fire chief in 1971, with the retirement of Russell C. Bowers.

The Reading Fire Department's staff of chiefs and fire prevention officers in the late 1960s includes, from left to right, Deputy Chiefs Harry Stauffer, John Weinhold, Merle Gerlach, Russell Mogel, Fire Chief Russell Bowers, Fire Marshal Ralph Pennypacker, Deputy Fire Chief Albert Batastini, and Deputy Fire Marshal Florin Monasmith.

Pictured here from left to right are Chiefs Harry Stauffer, John Weinhold, Merle Gerlach, and Russel Mogel as they take delivery of their new station wagons at the Lindgen Chrysler-Plymouth dealership on Lancaster Avenue around 1970.

The Liberty Fire Company's 1965 GMC/Hahn 1,000-gallon pumper became the first closed-cab engine bought by the city after the Union Fire Company paid for a cab on its 1949 Mack. In keeping with new apparatus standards, all apparatus purchased from then on had enclosed crew cabs. This engine departed from the Liberty's traditional all-white apparatus, using a white-over-red color scheme.

The fire department scuba team was organized by Deputy Chief Al Batastini in 1961 at the Friendship Fire Company, but its base of operation eventually moved to the Liberty company at Fifth and Laurel Streets. The team's original response vehicle was this 1958 white-over-red International/Reading body utility truck, previously used by the water department. This truck was finally replaced in 2002 with a bigger 1972 Ford rescue truck, formerly used by the Birdsboro Fire Department.

A fire in the huge Penn Hardware complex along the city's waterfront at Canal and Spruce Streets lights up the nighttime sky on March 14, 1971. A master stream from the Reading Hose Snorkel deluges part of the burning building, while a ladder pipe stream from the Schuylkill Fire Company's 75-foot Pirsch aerial blitzes the other end.

Firefighting crews take a well-earned coffee break on March 15, 1971, the day after the all-night fire that destroyed the Penn Hardware complex at Water and Spruce Streets. The Marion Fire Company's 1948 85-foot Peter Pirsch aerial ladder continues to apply water on hot spots with its ladder pipe stream. To the right is the Friendship Fire Company's 1949 Mack 750-gallon pumper, and behind the firefighters is the Neversink Fire Company's 1940 Mack 750-gallon engine.

The Marion Fire Company's 1948 Pirsch 85-foot ladder goes into service with a ladder pipe at a four-story rag and scrap warehouse in the 100 block of Maple Street on April 21, 1973.

As attendees to the Pennsylvania State Firemen's Convention in September 1973 watched the fire loom up from their Abraham Lincoln Hotel rooms, firefighters fought these garage fires, with vehicles in them, at Front and Buttonwood Streets. Career firefighter William Shunk (left) calls for water in the supply line to his 1959 Mack 750-gallon engine (with the civil unrest protective cage installed).

The Schuylkill's 1956 Pirsch ladder and the Reading Hose Pitman Snorkel perform aerial work at a row of burning homes in the 400 block of South Seventh Street on January 24, 1972.

After several smaller fires, the abandoned Reading Railroad Company freight station at Eighth and Buttonwood Streets is destroyed in this spectacular two-alarm fire on October 20, 1980.

Deputy Chief Earl R. Bansner takes a well-deserved coffee break during a three-alarm fire at the Fink Apartments on Fifth and Walnut Streets on November 18, 1979.

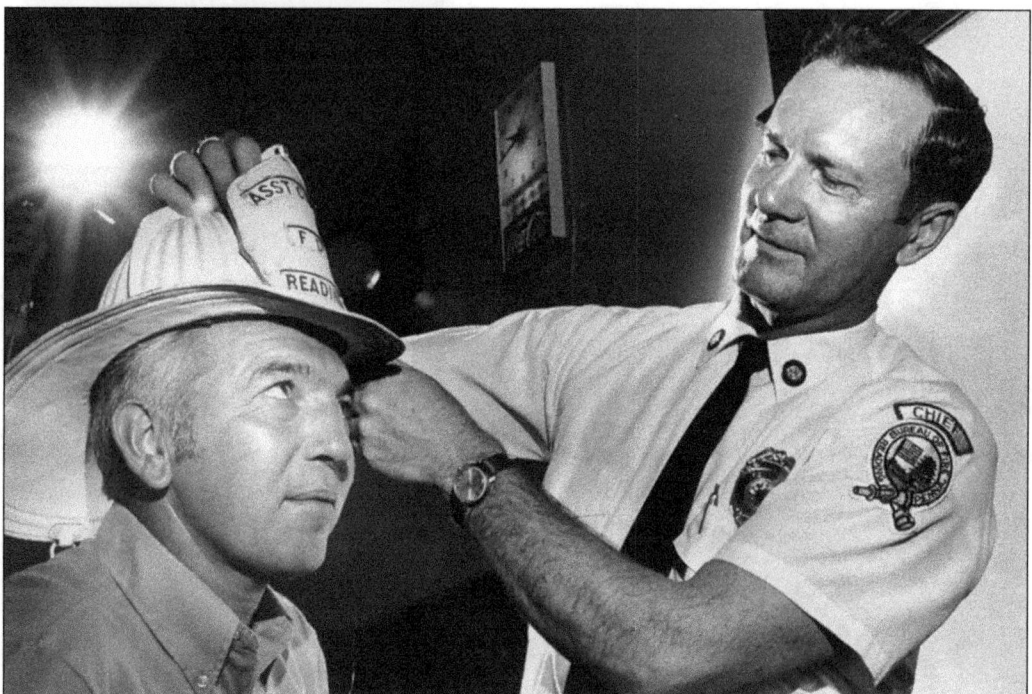

Fire Chief Russell P. Mogel presents Charles "Swifty" Schaeffer with his new helmet upon his promotion to assistant chief (later redesignated deputy chief) in July 1971.

The Reading Hose Snorkel backs into position alongside the Marion Fire Company's 1948 Mack 750-gallon pumper at a fire at the Children's Home, on 1010 Centre Avenue. A disgruntled juvenile started the fire in the north wing of the building.

An early-morning fire damaged the Ennis Tool Company at Eighth and Oley Streets on January 28, 1969. The fire was handled by a still-alarm assignment (two engines and two ladders), consisting of, from left to right, the Hampden's 1958 Mack, the Schuylkill's 1956 Pirsch ladder, the Marion's 1948 Mack, and the Reading Hose's 1968 Pitman Snorkel. (Note the price of gasoline at the Texaco gas station, which was 33.9¢ per gallon.)

Nine civilians in two families died when a gas explosion followed by fire, caused by a city water crew digging up the street for a broken water line, leveled two semi-detached homes at 49 and 51 North Carroll Street in the city's Oakbrook section on January 9, 1968. All of the victims were buried in the collapse.

The Reading Hose's 75-foot Pitman Snorkel prepares for deluge gun operations as civilians help drag fire hoses at the rag and scrap warehouse in the 100 block of Maple Street on April 21, 1973. The B'nai B'rith high-rise apartment building now occupies this spot.

The four-story Heller Bookbindery building as 45 Reed Street in downtown Reading was gutted by a windswept fire on Sunday, January 28, 1967. This unique view taken from the Berks County Courthouse shows the Junior Fire Company's 1959 Mack 750-gallon engine supplying a two-and-a-half-inch hand line and the Keystone's 1959 Mack-Maxim 85-foot aerial ladder pipe. Opposite the Junior's engine, the Neversink's 1940 Mack 750-gallon pumper feeds numerous hose lines along the south side of the building.

A victim is rescued from the wreckage of his row home in the 1400 block of Cotton Street in the East Reading section of the city following a gas explosion and fire, which leveled several homes and killed two others on January 22, 1978. Medical crews were ready to amputate the victim's foot, which was pinned by heavy structural debris, when firefighters were finally able to free him. The early-evening explosion followed an all-day blizzard that dumped more than a foot and a half of snow on the city.

The rear portion of the Harold's Furniture warehouse at Eighth and Green Streets is heavily damaged after lightning struck the building on June 22, 1979. The Rainbow Fire Company's 1963 Hahn/Diamond-T 1,000-gallon pumper pumps in the foreground, while to the left, the Schuylkill Fire Company's 1956 Pirsch aerial operates a ladder pipe. The entire building completely burned down on January 3, 1981, apparently the work of an arsonist.

Streaks of lightning in the eastern sky, over the pagoda on Mount Penn, are the remnants of the thunderstorm that caused the fire at the Harold's Furniture warehouse, on Eighth and Green Streets, on June 22, 1979. In this photograph, the burning warehouse can be seen behind the Reading Railroad outer station, viewed from the North Fifth Street bridge.

Three of Reading's ladder companies are in operation at this early-morning center-city fire at Rose and Walnut Streets on October 20, 1970, at a mixed-occupancy building. In the front on Walnut Street, the Neversink Fire Company's 1959 Mack 750-gallon engine works with the Keystone's 85-foot 1959 Mack/Maxim aerial ladder, while on Rose Street the Reading Hose's 1967 75-foot International/Pirsch Snorkel and the Schuylkill's 1956 75-foot Pirsch ladder use their aerials on the fire building and the exposures.

The Keystone Fire Company's 1959 Mack/Maxim 85-foot aerial ladder is used to apply an elevated master stream, while firefighters manning two-and-half-inch hose lines protect exposures on both sides of a wind-driven two-alarm fire heavily involving the Hertz Enterprises Warehouse at 441 North Front Street around 6:00 p.m. on April 20, 1981. The building involved is identical to the one on the left, which survived only to burn to the ground a few years later in a midnight fire. Ten row homes in the background, on the 400 block of Pear Street, were damaged by radiant heat from this fire.

The broken fire hydrant in the foreground spelled the demise of the A. T. Consoli Construction Company at 1240 Chester Street in the city's Glenside section on December 4, 1972, at 9:40 a.m. Heavy equipment from the construction company apparently damaged the hydrant, but it was never reported to the water authority. It was discovered when the first-arriving engine attempted to connect to it with a soft sleeve section of hose. The time needed to connect to other sparsely distributed hydrants in the area doomed the block-long building.

Firefighters wait for water in a two-and-a-half-inch hose line to attack a two-alarm fire at the A. T. Consoli Construction Company, on Lehigh and Chester Streets, in the city's Glenside section, on December 4, 1972. The closest hydrant, across the street from the company, had apparently been damaged by the company's heavy construction equipment and never reported. It was completely inoperative and resulted in delays in getting water. The building shown was completely destroyed.

Numerous tenants of the downtown Fink Apartments at Fifth and Walnut Streets were rescued by ladders and the Snorkel's platform during an early-morning fire on November 18, 1979, at 4:00 a.m. The fire started in the basement below the corner restaurant and spread through a central light well in the middle of the building to all four floors of the structure. In the foreground is the Neversink Fire Company's white-over-lime-yellow 1976 Hamerly 1,250-gallon pumper, while to the right the Reading Hose's 1967 International/Pirsch 75-foot Snorkel prepares to cradle its platform. To the far left is the Marion Fire Company's Pirsch 85-foot aerial, whose 1948 tractor had been replaced with a used 1966 International cab.

Deputy Fire Chief William H. Rehr checks the front of the Fink Apartment building on the southeast corner of Fifth and Walnuts Streets where smoke still issues from the basement after an all-night fire on November 18, 1979. Many tenants were rescued by ground ladders and aerial trucks.

Smoke still emanates from the Fink Apartment building at Fifth and Walnut Streets after an all-night fire gutted the building on November 18, 1979. Numerous tenants were rescued by ground and aerial ladders after the fire, spreading from the basement, cut off escape by the interior stairways. The Hampden Fire Company's 1958 Mack engine and the Schuylkill's 75-foot 1956 Pirsch aerial operate nose to nose in the 500 block of Walnut Street.

The McDonald's restaurant in the heart of downtown Reading, at the corner of Fifth and Penn Streets, was destroyed in this general-alarm fire on April 3, 1977. The Reading Hose's Pitman Snorkel (center), the Keystone's Mack/Maxim ladder (left), and the Washington's Hahn/Pirsch ladder (right) deploy elevated master streams along the South Fifth Street side of the restaurant/office complex.

Career firefighter Bruce Briner and a Liberty Fire Company volunteer set up a deluge gun across the street from the burning Textile Chemical Company at 122 Chestnut Street on January 18, 1972.

A Liberty Fire Company volunteer directs a 1.5-inch hose stream at a burning Textile Chemical Company building at 122 Chestnut Street on a warm winter afternoon, January 18, 1972. After the fire, the industrial chemicals that were contaminated were disposed of by a cleanup company in a dump on Neversink Mountain. About one year later, the chemicals reacted and started a hazardous materials dump fire that burned for more than two years.

The Carpenter Steel Company fell victim to another costly fire when this blaze heavily damaged building No. 48, on the company's West Shore facilities in the city's Glenside section on May 26, 1977. The fire burned off most of the roof covering of the two-block-long wire mill and damaged rooftop equipment and ventilation systems.

Supervisory personnel from the Carpenter Steel Company discuss strategy for resuming production after a fast-moving fire heavily damaged the roof and equipment of the two-block-long building No. 48, on May 26, 1977. The Keystone Fire Company's 1959 tractor-drawn 85-foot Mack/Maxim ladder is parked in the background.

Deputy Fire Chief Merle A. Gerlach served as a deputy chief from 1968 until his retirement in 1978. Gerlach was a longtime member of the Junior Fire Company and worked as a paid driver at the Rainbow Fire Company before being appointed deputy chief.

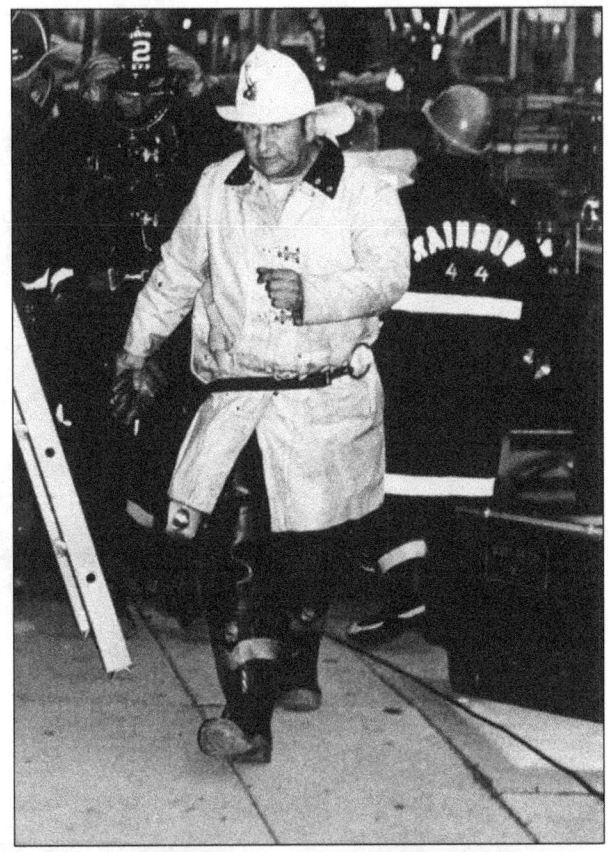

The Reading Hose's 1968 Pitman Snorkel makes its first Labor Day debut on Penn Square in 1968. This photograph was taken through the International cab windshield of the Washington Fire Company's 100-foot aerial truck. The Union Fire Company's 1949 closed-cab 750-gallon Mack pumper passes on the right.

Cadillac ambulances were popular throughout the 1950s, 1960s, and 1970s, when they were bought by the United Fund (forerunner of the United Way) and titled to the three city fire companies that operated ambulances. The Junior Fire Company's unit is pictured here in front of its Reed and Walnut Streets station.

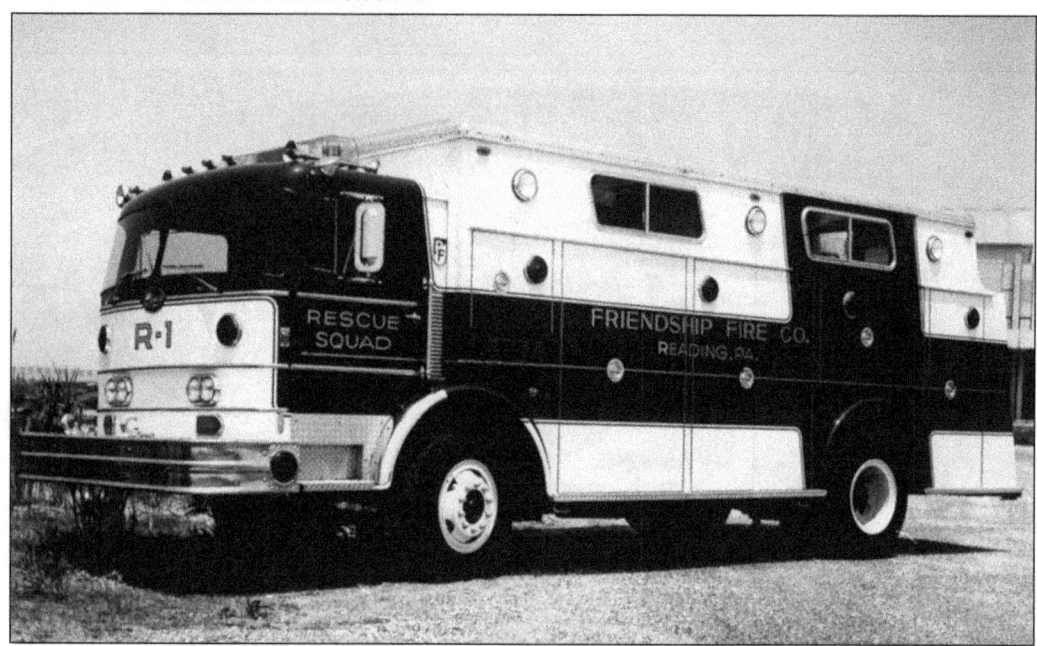

The Friendship Fire Company's third-generation rescue truck, which replaced a 1957 GMC unit, was this 1976 Hamerly vehicle. The city also bought three Hamerly 1,250-gallon pumpers in the same year. This rescue truck served until it was replaced in 1993 by a Mack/Saulsbury rescue truck.

The Schuylkill Fire Company's 1956 Pirsch 75-foot aerial extends to the third-floor dormer window of a dwelling during a one-alarm fire in the 300 block of Heller's Court on April 25, 1966. The extremely narrow dead-end street consisted of a row of wood-frame homes, which made fireground operations difficult. They were torn down during urban renewal in the 1970s.

The Keystone's Mack/Maxim ladder and the Reading Hose Pitman Snorkel operate nose to nose at the general-alarm fire that destroyed the McDonald's restaurant and business offices at Fifth and Penn Streets in the heart of Reading on April 3, 1977. The view in the photograph is looking south on Fifth Street from Penn Street.

Firefighters pour water from two hand lines into the open fronts of 600 and 602 Fern Avenue after the front walls were blown out by a natural gas explosion. The husband and wife occupants of 602 Fern Avenue were both killed. Several firemen were injured when a second explosion occurred just after the companies arrived.

A natural gas explosion caused by a broken curb fitting destroyed three homes at the corner of Fern Avenue and Noble Street on Sunday afternoon, February 17, 1963. The husband and wife occupants of 602 Fern Avenue were both killed. Five years later, nine people died in another gas explosion a few blocks away.

The Reading Hose 75-foot Pitman Snorkel (right) applies a deluge stream from the bucket, at its position in front of the building, while volunteers man a two-and-a-half-inch hose line along the north side of the Reading Railroad outer station on February 20, 1978. The illuminated tops of the Berks County Courthouse and the steeple of the Trinity Lutheran Church in downtown Reading can be seen under the Snorkel's boom.

Firefighters standing in deep snow and ice apply a two-and-a-half-inch hose stream to the fully involved Reading Railroad outer station depot in the 500 block of North Sixth Street on February 20, 1978. Snow and ice accumulations hindered the deployment of apparatus and the laying of hose lines.

A smoking shell is all that remains of the 100-year-old Reading Railroad outer station depot off the 500 block on North Sixth Street after an all-night fire destroyed the building on February 20, 1978. Driveways up to the building were clogged with heavy snow and ice accumulations, making access for apparatus difficult. The Marion Fire Company's tractor-drawn aerial was stuck in the snow and had to be towed out after the fire was out.

The rear entrance to the Reading Railroad outer station is covered with icicles the morning after a frigid all-night fire destroyed the 100-year-old passenger depot on February 20, 1978.

A two-alarm fire heavily damaged the Hodges Apartment building on the northeast corner of Fifth and Greenwich Streets on January 9, 1980. Another two-alarm fire broke out in the complex on January 24, 1980. Both fires were apparently the work of an arsonist.

A painter using a blowtorch at the base of one of the white fluted columns started a fire inside the column, which spread up to the attic overhang in the front of the Reich Apartments at Good and Clymer Streets on June 13, 1979. The fire burned off the roof and gutted the attic.

The Washington Fire Company's 100-foot aerial attempts to stop the progress of the fire burning the roof off the Textile Chemical Company plant at 122 Chestnut Street on January 18, 1972.

The Washington Fire Company's 100-foot International/Pirsch aerial operates a ladder pipe amid drums of volatile and corrosive materials at a two-alarm fire at the Textile Chemical Company plant at 122 Chestnut Street on January 18, 1972.

Firefighters man two-and-a-half-inch hose lines in an attack on a smoky two-alarm fire at the Eways Rug Company at Wood and Cherry Streets in downtown Reading on May 6, 1963. The fire involved large amounts of carpeting and foam underlay material.

The Marion Fire Company's International/Pirsch 85-foot aerial deluges the burning Reading Railroad outer station at the 500 block on North Sixth Street on February 20, 1978, as the fire consumes the roof and clock tower. The truck was positioned at the rear of the building after struggling through heavy snow and ice but had to be towed out after the fire was extinguished.

The Schuylkill Fire Company's 1956 Pirsch 75-foot aerial is used to apply an elevated master stream on the burning demolition debris of the House of Good Shepherd, 1130 Schuylkill Avenue, in the early 1970s. Wrecking company employees using a cutting torch accidentally started the fire. Operator of the aerial career firefighter George Hunsinger (left) converses with Liberty Fire Company volunteer firefighter John Kissinger.

Hose lines clog Carpenter Street at the Reading Merchants Oil Company warehouse between Cherry and Franklin Streets on January 16, 1970. The empty hose bed (except for the plywood platform above it) of the Neversink Fire Company's 1959 Mack pumper in the background undoubtedly contributed to much of the hose in the street.

The Reading Merchants Oil Company warehouse, at 33 Carpenter Street in downtown Reading, burns on January 16, 1970. In the foreground, the Washington Fire Company's 1951 Mack 750-gallon pumper operates from a hydrant at Fourth and Cherry Streets, while the Reading Hose's Snorkel, in the background, operates a deluge stream from its bucket at Carpenter and Cherry Streets.

The Keystone Fire Company's 1959 Mack/Maxim tractor-drawn aerial operates a ladder pipe on the burning Redcay Produce Company in the 300 block on North Third Street on September 16, 1979. In front of the ladder, the Reading Hose's Pitman Snorkel prepares for a deluge nozzle operation.

The Washington (left) and Keystone (right) aerials prepare for ladder pipe operations at the Penn Hardware complex at Canal and Spruce Streets on September 11, 1961. Two suburban volunteer firemen, who were employees in one of the businesses, were later arrested after they set several more fires. The entire complex burned down in a spectacular fire in March 1971.

Assistant Fire Chief John Weinhold (foreground) supervises mop-up operations after an all-night, two-alarm fire at the Luden Candy Company warehouse at 628 South Ninth Street on July 24, 1971. The fire destroyed large quantities of Luden's candy packaging materials. The Liberty Fire Company's 1965 GMC/Hahn pumps numerous hose lines while the Reading Hose's Snorkel bucket works the roofline.

The Union Fire Company's 1949 closed-cab Mack 750-gallon pumper was responding to an automobile fire when it collided with a civilian vehicle at Eleventh and Penn Streets, where Perkiomen Avenue begins. Both vehicles mounted the sidewalk, knocked down a traffic signal, and struck the front of the building at 1100 Perkiomen Avenue. The driver of the car was injured in the December 1973 accident.

City firefighters conduct high-angle rescue operations after a car plunged over the wall at the parking outlook at the pagoda on Skyline Drive on April 20, 1961. The driver of the convertible (sitting on the right) accidentally had the car in a forward gear instead of reverse when she went to back out of a parking space. Her friend in the passenger seat was crushed and killed.

A trash hauler delivered a load of burning rubbish to the Liberty Fire Company station at Fifth and Laurel Streets on March 26, 1975. Two volunteers in somewhat casual attire wet down the rubbish with a one-and-a-half-inch hose line.

A ladder pipe is deployed from the Keystone Fire Company's 1959 Mack/Maxim 85-foot tractor-drawn ladder at the Hertz Supply Company warehouse at Front and Buttonwood Streets on April 20, 1981. The windswept fire spread to several homes nearby in the 400 block of Pear Street.

Three aerials, seen here from left to right, the Washington's 100-foot Hahn/Pirsch, the 75-foot Reading Hose's Pitman Snorkel, and the 85-foot Keystone's Mack/Maxim, are engaged in mop-up operations at the Park Theatre-Daniel Boone Hotel complex in the 1000 block of Penn Street, after an all-night fire on May 20, 1978.

Even fire trucks catch fire, as evidenced by the smoke pouring from the Washington Fire Company's 1975 Hahn/Pirsch 100-foot aerial that was parked temporarily in the Rainbow Fire Company's station around 1980. The malfunction also prevented the truck from being shut off until the exhaust pipe was blocked by a firefighter, causing the engine to stall.

Firefighters raise a 35-foot extension ladder on the Bingaman Street side of the Crescent Brass Foundry at Seventh and Bingaman Streets on May 20, 1991. The fire went to two alarms and destroyed most of the building.

A one-and-one (one engine and one ladder) still-alarm assignment was sent on a snowy night, December 12, 1989, to the East Penn Refractories Company at 1250 Clarion Street in the city's Glenside section for an automatic fire alarm. Upon arrival, the crews of the Schuylkill Fire Company's 750-gallon GMC engine and the Marion's 85-foot KME/Maxim ladder (both shown in the photograph), found a working fire and struck the box. The fire eventually went to a second alarm as firefighters struggled with increasingly worse weather conditions.

The Washington Fire Company's 1975 Hahn/Pirsch 100-foot aerial gets into position to set up a ladder pipe to protect the Prizer-Painter manufacturing complex, adjacent to the burning Marketing Specialists warehouse at 620 Arlington Street, on April 21, 1984. The Marketing Specialists building was destroyed, but the Prizer-Painter buildings were saved.

Career firefighter Kenneth Moyer, operator of Ladder 1 (formerly Keystone Fire Company), brings a child off the turntable after she was carried down the aerial ladder from the third floor of an apartment house fire at Front and Walnut Streets on November 18, 1989.

Fire gutted this corner end of row homes at Printz and Muhlenberg Streets during the evening of February 15, 1985. The two narrow streets in the East Reading section of the city precluded most apparatus from entering and required hose lines to be stretched from nearby Cotton Street.

Deputy Fire Chief William H. Rehr assumes command of a working second-floor fire in a row home at 419 Maple Street in southeast Reading on February 23, 1991. The fire heavily damaged the home and burned off electrical service in the front of the building.

The Reading Grey Iron Company, at Tulpehocken and West Green Streets, fell victim to many incendiary fires after it was shut down. This fire on January 28, 1992, was the final one prior to demolition.

This Snorkel prepares to go into operation at a two-alarm fire at the American Chain and Cable complex on October 10, 1994. It was one of many arson fires that were set in the sprawling complex of old industrial buildings at Tulpehocken Street and the Lebanon Valley Railroad.

Deputy Fire Chief William H. Rehr investigates an early-morning fire that gutted a barroom at Eleventh and Green Streets on December 7, 1978. A woman was rescued from a third-floor window ledge after smoke cut off her inside escape route. Rehr went on to become fire chief in 1996.

Reading firefighters apply acid-neutralizing foam on a leaking 10,000-gallon hydrofluoric acid tank at the Carpenter Steel Company on August 9, 1989. Firefighters and plant fire brigade personnel eventually secured the leak.

Reading's famous landmark, the pagoda, located on top of Mount Penn has survived several fires during its lifetime. This fire on a warm, windy evening in the late 1970s started in the neon lighting on the city side of the entrance balcony and wrapped around to the entrance side. Most of the fire was contained to the wooden underside of the balcony roof overhang. The first-due Union Fire Company engine, a 1976 Hamerly 1,250-gallon pumper, provided hand lines to extinguish the fire, while the Washington Fire Company's 100-foot 1975 Hahn/Pirsch ladder has its main raised to the third floor.

A firefighter, balanced on the edge of the balcony railing with the aid of another firefighter holding on to his bunker coat, finishes wetting down the underside of the balcony roof overhang at Reading's famous pagoda on Skyline Drive. The fire began in neon lighting used to outline each roof level of the building.

Visit us at
arcadiapublishing.com

www.ingramcontent.com/pod-product-compliance
Lightning Source LLC
Chambersburg PA
CBHW081419160426
42813CB00087B/2249